In Quest of The Quark

In Quest of The Quark

A Student's Introduction to Elementary Particle Physics

By

Dr. Linda Bartrom-Olsen

Graphic Artist
Juan Pablo Larios

Library of Congress Control Number:		2013922217
ISBN:	Hardcover	978-1-4931-5084-7
	Softcover	978-1-4931-5083-0
	eBook	978-1-4931-5085-4

This book was printed in the United States of America.

Rev. date: 01/10/2014

To order additional copies of this book, contact:
Xlibris LLC
1-888-795-4274
www.Xlibris.com
Orders@Xlibris.com
140486

CONTENTS

LIST OF FIGURES

LIST OF TABLES

DEDICATION

In memory of Eric
"from a small family on a big street"
who is in my thoughts and in my heart
every moment of my life.

. . . . and to my dad
whose nights at the kitchen table
with me and my math
live forever in my heart.

Appreciation to Rotary International
and Rotary District 5320 for their
support of this cross-cultural project.

"Everything comes from everything.
Everything is made from everything,
And everything can be turned
Into everything else."

Leonardo de Vinci

CHAPTER 1

The Beginning

A S LONG AGO as the Greek age of Aristotle, man has been proposing that the chemical make-up of his world was based on the tiny unit called the "atom" from the Greek word *atomos* meaning indivisible (or not dividable). As chemical theory progressed through the ages, physicists began to make inquiries into the interior composition of the atomic unit. This resulted in the Periodic Table of the Elements designed by Mendeleev during the 19th century. Finally, in the 20th century the truth was discovered: not only is the atom dividable, but it is composed of precisely arranged, discretely categorizable particles.

The classic Atomic Theory portrait bases matter on a unit known as the atom, composed of a centrally located nucleus which contains various numbers of protons and orbited (at near light speed) by various numbers of electrons. The smallest atom is that of Hydrogen with only one proton, positively charged, as the nucleus, and orbited by one negatively charged electron. The charges then cancel each other and the atom is neutral. All other atoms are multiples of Hydrogen, from Helium with two protons and two electrons, to uranium with 92 protons orbited by 92 electrons shown in **Figure 1.** Additionally, neutrally charged particles called neutrons may also reside in the nucleus. Since neutrons have no charge, they do not disturb the neutrality of the atom, but only add to its mass. The larger atoms with greater numbers of electrons surrounding their nuclei have these electrons arranged in specific levels of orbital paths. It is, however, only the number of *outermost* electrons that determine an atom's physical and chemical characteristics, including whether or not, how, and to what extent, the atom will interact with the outermost electrons of another atom.

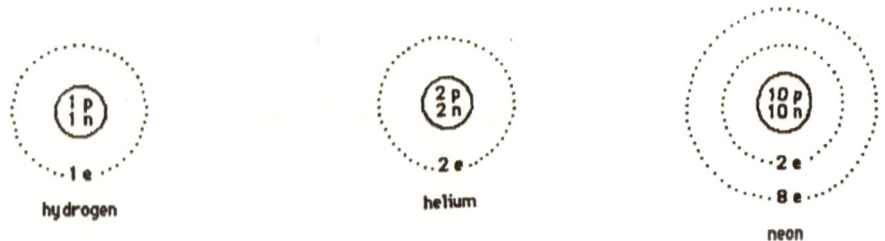

Figure 1 H, He, and Ne

As time progressed, other particles made their appearance on the list of sub-atomic particles. Although their charges and masses (or lack of it) could be detected, their relationships to each other remained, for the most part, an anomaly; until the 1960's that is. While the political world was rending itself apart from within, and the hallowed halls of higher learning were becoming hotbeds of dissension as college campuses entered The Age of the Demonstration, a quieter revolution was occurring in scientific circles: evidence had finally been amassed for formal proposals concerning previously unknown particles of an ordered nature that actually *make up* subatomic particles: a family of sub-subatomic particles, as it were. At first, these constituent subatomic particles were divided into two groups: leptons and hadrons.

One of the prime revisions in the outlook required for understanding the investigation of sub-subatomic particles, is the suspension of what is generally referred to as *Newtonian Physics*. Within this classic theory of physics, matter was designated as having mass-location as a function of three space dimensions and (generally) one time dimension. The study of particles as small as *leptons*, or the *quarks* which compose *hadrons* (like protons and neutrons), and which move within their incredibly tiny domains

DR. LINDA BARTROM-OLSEN

at speeds far beyond our comprehension, necessitates a revision in outlook. What is now termed *The Heisenberg Uncertainty Principle*, named for its author, addresses itself to the problem.

Essentially, *The Heisenberg Uncertainty Principle* concludes that as the momentum of a particle increases, the probability of predicting the location of (and therefore enabling the study of) its mass decreases. Thus, the new physics of sub-subatomic particles, now termed Elementary Particle Physics, relies heavily on probability, statistical averages, and redefining of the particle-mass theory itself in terms of "rest-energy." Thereby was initiated a physics of *"matter/energy"* which Einstein had predicted, and upon which thermodynamics had been based in theory for a number of years.

Despite the use of unfamiliar names, the particles themselves which comprise the world of sub-atomic particles, display an amazingly consistent behavior within their respective groups. Their behaviors, however, it must be remembered (such as in Quark Theory) are merely *models* which presently satisfy known data, and which are subject to change as the technology used in Elementary Particle Physics provides additional information. The breakdown of many of the heavier hadrons, for example, is quite predictable and analogous to nuclear degeneration in radioactive decay series. Their names could be listed here, but terminology is not the point. Terms for sub-atomic particles are only of value insofar as *Elementary Particles Physics* and *Quark Theory* enable us to understand that "stuff" called "matter", and to comprehend the "energy" which brings about changes in it. The prime and simplistic subdivisions within *Quark Theory* of *fermions* (analogous to matter) and *bosons* (analogous to *energy*) plus leptons and hadrons, are designed to do just that. But before we approach sub-subatomic particles, first let's take a look at the ongoing quest over the last two centuries to organize and make sense of the growing number of elementary particles called atoms.

CHAPTER 2

The Quest for Simplicity: History of the Development of The Periodic Table of the Elements

SCIENCE HAS LONG and often used graphics to interrelate concepts; the periodic table is one of the more profound examples of advance organizers. What better example of a structural graphic can there possibly be than The Periodic Table of the Elements? It is a concept map of the highest level, interrelating today the characteristics of over one hundred atoms into a meaningful structure.

Evolving over time, elements were organized by weight, and then arranged according to behavior, delineating the regularity of behaviors both physical and chemical, and paving the way to understanding why those properties existed. As the number of elements grew, the standard periodic table left lanthanides and actinides suspended at the bottom, rendering their relationship to the rest enigmatic. An expanded periodic table of the elements which interrelates all the elements with each other presents a useful first graphic in chemistry. This is of particular importance for high school students where unsophisticated learners are introduced to the complex array of over one hundred elements currently included in the table of elements.

From the early concept in B.C. of Earth, Air, Fire and Water, the early Greeks called the particles *atomos*, i.e. indivisable, (note—*not* "invisible"), because they thought each was a little solid particle like a "b-b". The Greek elements sometimes called the "-ons", like carbon, hydrogen, boron, nitrogen, and oxygen for example, were the ones the Greeks designated as being elemental. As elements increased in number they were given names and listed by weight, until the past century and a half saw the Periodic Table as we know it today begin to emerge. As the number of atoms increased, and as a clearer understanding of relationships emerged, they were listed by weight.

By 1865 Newland had organized elements by behavior through what he called his "Law of Octaves", in 1869 Dmitri Mendeleev's "Periodic System" was published, and by mid-twentieth century sub-orbital categories gave rise to today's periodic table. All three are examples of the steadfast insistence of the scientific mind, as the number of elements increased, to rearrange them in an effort to emphasize their interrelationships. These graphic organizers played no small part in the emerging insight into behavioral groupings and the search for the reason behind those behaviors.

The move in the mid-1800's from lists of elements by weight to the creative insight that their behavioral relationships could be organized into a structure that made these relationships obvious, was a bolt of lightning to chemistry. Largely unrecognized in his time, John Alexander Reina Newlands wrote in *Chemical News in 1865*:

> If the elements are arranged in order of their equivalents [*ie* relative atomic masses in today's terminology] with a few transpositions, it will be seen that elements belonging to the same group appear in the same horizontal line. Also the numbers of similar elements differ by seven or multiples of seven. Members stand to each other in the same relation as the extremities of one or more octaves of music . . . . This peculiar relationship I propose to call The Law of Octaves.

The Newlands table, **Figure 2**, was based on chemical behaviors repeating every eight elements; interestingly, it does something

Mendeleev's table did not: Newlands assigned *numbers* to the elements, not their weights. The lowest weights of different behaviors are arranged vertically at left, with elements which act like them going across next to them. The concept structure of the elements had begun, clearly postulating through his graphic a prediction of patterns in the properties of the elements.

H 1	F 8	Cl 15	Co/Ni 22	Br 29	Pd 36	I 42	Pt/Ir 50
Li 2	Na 9	K 16	Cu 23	Rb 30	Ag 37	Cs 44	Tl 53
Gl 3	Mg 10	Ca 17	Zn 25	Sr 31	Cd 34	Ba/V 45	Pb 54
Bo 4	Al 11	Cr 18	Y 24	Ce/La 33	U 40	Ta 46	Th 56
C 5	Si 12	Ti 19	In 26	Zr 32	Sn 39	W 47	Hg 52
N 6	P 13	Mn 20	As 27	Di/Mo 34	Sb 41	Nb 48	Bi 55
O 7	S 14	Fe 21	Se 28	Ro/Ru 35	Te 43	Au 49	Os 51

Figure 2 Newlands' Table

Four years after Newlands proposed his Law of Octaves and the table that shows it, Dmitri Mendeleev presented his "periodic system" to the Russian Chemical Society on March 6, 1869, and published it the same year in *Zeitschrift für Chemie*. He is remembered as the creator of the "Periodic Table" although he called it a "periodic system" **Figure 3.** He included all the known elements, and trusted his system enough to leave gaps where the structure logically predicted an atom should be. The way he developed the graphic was to write the properties of elements on pieces of card and (tradition has it) after organizing the cards he suddenly realized that if he arranged the element cards in order of increasing atomic weight, types of elemental behavior occurred repeatedly or periodically. Thus his name for the structure: "Periodic System."

DR. LINDA BARTROM-OLSEN

			Ti = 50	Zr = 90	? = 180
			V = 51	Nb = 94	Ta = 182
			Cr = 52	Mo = 96	W = 186
			Mn = 55	Rb = 104,4	Pt = 197,4
			Fe = 56	Ru = 104,4	Ir = 198
		Ni = Co = 59		Pd = 106,6	Os = 199
H = 1			Cu = 63,4	Ag = 108	Hg = 200
	Be = 9,4	Mg = 24	Zn = 65,2	Cd = 112	
	B = 11	Al = 27,4	? = 68	Ur = 116	Au = 197?
	C = 12	Si = 28	? = 70	Sn = 118	
	N = 14	P = 31	As = 75	Sb = 122	Bi = 210?
	O = 16	S = 32	Se = 79,4	Te = 128?	
	F = 19	Cl = 35,5	Br = 80	J = 127	
Li = 7 Na = 23		K = 39	Rb = 85,4	Cs = 133	Tl = 204
		Ca = 40	Sr = 87,6	Ba = 137	Pb = 207
		? = 45	Ce = 92		
		?Er = 56	La = 94		
		?Yt = 60	Di = 95		
		?In = 75,6	Th = 118?		

1. Die nach der Grösse des Atomgewichts geordneten Elemente zeigen eine stufenweise Abänderung in den Eigenschaften.

2. Chemisch-analoge Elemente haben entweder übereinstimmende Atom-gewichte (Pt, Ir, Os), oder letztere nehmen gleichviel zu (K, Rb, Cs).

3. Das Anordnen nach den Atomgewichten entspricht der *Werthigkeit* der Elemente und bis zu einem gewissen Grade der Verschiedenheit im chemischen Verhalten, z. B. Li, Be, B, C, N, O, F.

4. Die in der Natur verbreitetsten Elemente haben *kleine* Atomgewichte

Figure 3 Mendeleev's Periodic System

These two scientists discovered new information and made sense of it by linking concepts to other concepts. Further, their work was based solely on observation, the purest of empiricism. There was, in the mid 1800's, absolutely no reason to think that the elements would show periodicity, there was no preconceived idea guiding their work, and a reasonable explanation for the phenomenon would not be available for another 50 years. The work of Newlands and Mendeleev was astounding in its creativity and boldness.

The standard periodic table of the elements used today is familiar to the learner but much of the language representing concepts about periodicity is completely new. The table's squares are filled with a plethora of physical and chemical data and the organization is based on sub-orbitals with the 2-s electrons at the left and the 6-p electrons at the far right, completing the octet in the outermost shells. The middle is occupied by the drop-down of the 10 d-electron elements, the "transition metals" filling in two levels beneath the outermost, and then the lanthanides and actinides, the "rare earth" metals, filling shells three levels beneath the outermost. These latter elements are separated from the rest of the chart, down at the bottom in two rows.

The sheer number of atoms making up our universe is difficult for students to comprehend and made more enigmatic by the separation of the lanthanides and actinides off by themselves at the bottom of the chart. Interestingly, the fact that they are hanging off by themselves at the bottom of the paper or chart is largely due to the size and proportion of paper, not any pedagogical or theoretical consideration. If the squares are going to be large enough to hold information then this arrangement is the practical solution, **Figure 4**. But where does this leave the beginning students of chemistry?

DR. LINDA BARTROM-OLSEN

The Periodic Table of the Elements

1	2	3	4	5	6	7	8	9	10	11	12	13	14	15	16	17	18
1 **H** Hydrogen 1.00794																	2 **He** Helium 4.003
3 **Li** Lithium 6.941	4 **Be** Beryllium 9.012182											5 **B** Boron 10.811	6 **C** Carbon 12.0107	7 **N** Nitrogen 14.00674	8 **O** Oxygen 15.9994	9 **F** Fluorine 18.9984032	10 **Ne** Neon 20.1797
11 **Na** Sodium 22.989770	12 **Mg** Magnesium 24.3050											13 **Al** Aluminum 26.981538	14 **Si** Silicon 28.0855	15 **P** Phosphorus 30.973761	16 **S** Sulfur 32.066	17 **Cl** Chlorine 35.4527	18 **Ar** Argon 39.948
19 **K** Potassium 39.0983	20 **Ca** Calcium 40.078	21 **Sc** Scandium 44.955910	22 **Ti** Titanium 47.867	23 **V** Vanadium 50.9415	24 **Cr** Chromium 51.9961	25 **Mn** Manganese 54.938049	26 **Fe** Iron 55.845	27 **Co** Cobalt 58.933200	28 **Ni** Nickel 58.6934	29 **Cu** Copper 63.546	30 **Zn** Zinc 65.39	31 **Ga** Gallium 69.723	32 **Ge** Germanium 72.61	33 **As** Arsenic 74.92160	34 **Se** Selenium 78.96	35 **Br** Bromine 79.904	36 **Kr** Krypton 83.80
37 **Rb** Rubidium 85.4678	38 **Sr** Strontium 87.62	39 **Y** Yttrium 88.90585	40 **Zr** Zirconium 91.224	41 **Nb** Niobium 92.90638	42 **Mo** Molybdenum 95.94	43 **Tc** Technetium (98)	44 **Ru** Ruthenium 101.07	45 **Rh** Rhodium 102.90550	46 **Pd** Palladium 106.42	47 **Ag** Silver 107.8682	48 **Cd** Cadmium 112.411	49 **In** Indium 114.818	50 **Sn** Tin 118.710	51 **Sb** Antimony 121.760	52 **Te** Tellurium 127.60	53 **I** Iodine 126.90447	54 **Xe** Xenon 131.29
55 **Cs** Cesium 132.90545	56 **Ba** Barium 137.327	57 **La** Lanthanum 138.9055	72 **Hf** Hafnium 178.49	73 **Ta** Tantalum 180.9479	74 **W** Tungsten 183.84	75 **Re** Rhenium 186.207	76 **Os** Osmium 190.23	77 **Ir** Iridium 192.217	78 **Pt** Platinum 195.078	79 **Au** Gold 196.96655	80 **Hg** Mercury 200.59	81 **Tl** Thallium 204.3833	82 **Pb** Lead 207.2	83 **Bi** Bismuth 208.98038	84 **Po** Polonium (209)	85 **At** Astatine (210)	86 **Rn** Radon (222)
87 **Fr** Francium (223)	88 **Ra** Radium (226)	89 **Ac** Actinium (227)	104 **Rf** Rutherfordium (261)	105 **Db** Dubnium (262)	106 **Sg** Seaborgium (263)	107 **Bh** Bohrium (262)	108 **Hs** Hassium (265)	109 **Mt** Meitnerium (266)	110 (269)	111 (272)	112 (277)	113	114				

58 **Ce** Cerium 140.116	59 **Pr** Praseodymium 140.90765	60 **Nd** Neodymium 144.24	61 **Pm** Promethium (145)	62 **Sm** Samarium 150.36	63 **Eu** Europium 151.964	64 **Gd** Gadolinium 157.25	65 **Tb** Terbium 158.92534	66 **Dy** Dysprosium 162.50	67 **Ho** Holmium 164.93032	68 **Er** Erbium 167.26	69 **Tm** Thulium 168.93421	70 **Yb** Ytterbium 173.04	71 **Lu** Lutetium 174.967
90 **Th** Thorium 232.0381	91 **Pa** Protactinium 231.03588	92 **U** Uranium 238.0289	93 **Np** Neptunium (237)	94 **Pu** Plutonium (244)	95 **Am** Americium (243)	96 **Cm** Curium (247)	97 **Bk** Berkelium (247)	98 **Cf** Californium (251)	99 **Es** Einsteinium (252)	100 **Fm** Fermium (257)	101 **Md** Mendelevium (258)	102 **No** Nobelium (259)	103 **Lr** Lawrencium (262)

Figure 4 Standard Periodic Table

So as technology advanced and the number of elements has grown, the graphic of the standard periodic table of elements left (due to paper-proportion constraints) two rows of heavy elements hanging free at the bottom, as if "created later" my students sometimes say, and with their relationship to the rest unspecified. For the introduction of the elements, however, a structure which interrelates *all* elements as one integral unit, and incorporates the lanthanides and actinides could establish the unity of the table. While not showing behavioral details of the classic form due to smaller squares for each element, an expanded version clearly shows the novice learner that *all* elements are truly part of one whole. It reinforces an inclusive perception of elemental interrelationships before using the standard form for individual atomic details.

The expanded periodic table shows the novice learner that all the elements are truly part of one whole. This graphic can be used to introduce the elements; it conceptually interrelates *all* the elements and provides a solid more inclusive perception of elemental interrelationships for beginning chemistry students. The expanded periodic table shown in **Figure 5**, facilitates the introduction of elemental relationships in a unified form before using the standard form of the table with its larger squares and expansive information.

The Expanded Periodic Table of Elements

Developed by: Dr. Bartrom, VPHS
Layout by: Pablo Larios, Graphic Design VPHS

Figure 5 Expanded Periodic Table

DR. LINDA BARTROM-OLSEN

The periodic table of the elements provides access for learners to perceive the multiple interrelationships among atoms: atomic number, shells denoting size by row, electrons in the outermost shell denoting behavior by vertical column. The additional fitting into this structure of the actinides and lanthanides where they belong through the expanded table, although the squares have less detail, provides students a complete graphic organizer. This less detailed graphic provides a framework at the beginning of understanding, followed by using more detailed graphics, vis-à-vis the standard table with larger squares holding much more information, later. Logical structured graphics which emphasize of the interrelationships of concepts provide a learning tool during instruction, a cognitive structure for retention, and a basis for the acquisition of further knowledge. This was true developmentally in the history of chemistry as a science, and is just as true today in the classroom where we pass that legacy of knowledge forward into the future.

CHAPTER 3

The Composite Units of Matter/ Energy: Bosons and Fermions

MATTER AND ENERGY in Elementary Particle Physics move beyond the ideas of protons, neutrons and electrons, and settle into a more incisive family of terms which begin at the level of bosons and fermions. Fermions (analogous to matter) may be considered the "bricks" of which the universe is made, while bosons (analogous to energy) may be considered the "mortar" which binds them together. Fermions essentially interact with one another by exchanging bosons.

The prime difference between bosons and fermions is related to *The Pauli Exclusion Principle*, formulated in 1925 by Wolfgang Pauli. The Exclusion Principle states that no two particles of any one type may occupy the same quantum state (read location in space) at the same time, therefore electrons are spread out at probability levels, **Figure 6.**

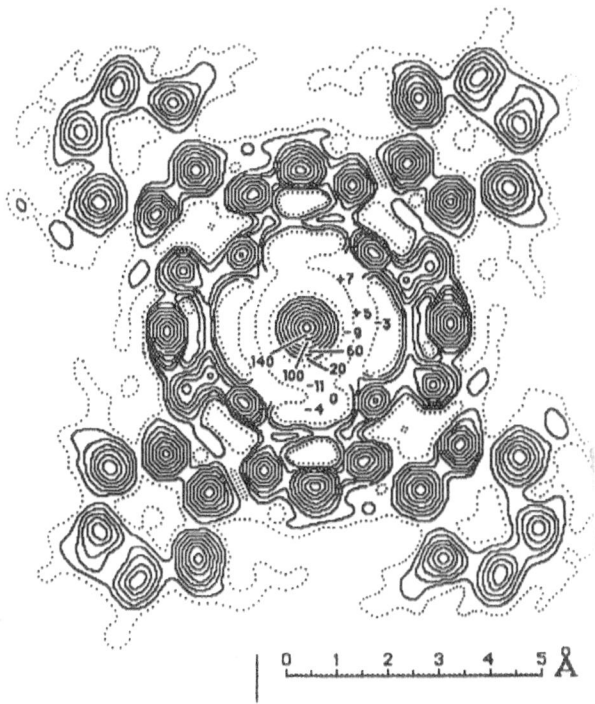

Figure 6 Electron Density Map

Fermions *are* affected by the Exclusion Principle and tend *not* to occupy the same space at the same time, i.e. they repulse each other, and each must exist "in its own space." Bosons on the other tine do *not* conform to the Exclusion Principle, and therefore they cluster. They may, in theory, travel in massive groups, with their *quantity* determining the *amount* of energy and the *type* of boson determining the *kind* of energy.

Bosons

Bosons are classified according to the kinds of changes they bring about on fermions (matter), and these fundamental classifications of change are called the four Forces. The **Strong Force**, which binds protons and neutrons together is controlled by a boson family called **gluons** by elementary particle physicists.

The **Electromagnetic Force**, which binds oppositely charged particles and produces the Electromagnetic Spectrum of energy, is transmitted by clusters of bosons termed **photons.** The **Weak Force,** which is involved with nuclear decay and brings about change in the nature of particles, is transmitted by **intermediate vector bosons.** Finally, the **Gravitational Force** results, in theory, from the exchange of **bosons** called **gravitons** or "intermediate vector baseballs." A Higg's Field actuated by Higg's Boson has also been discovered with contention that it may play a part in gravitational force. **Table 1** organizes the Forces and the Bosons which bring about their effects.

Boson Characteristics: Particles of Energy
Table 1

Boson	Relative Strength	Force which Results	Symbol	Number	Spin Mass	Interact with	Function: To change...
Gluons	strongest	Strong Force	(none)	8		quarks	the "color" of the Quarks
Photons	strong	Electro-magnetic Force	γ	1	$1\hbar$	charged units	hadrons and atoms
Intermediate Vector Bosons	weak	Weak Force	$w^+, w^-, z^\circ,$	3	$1\hbar$	hadrons	the "flavor" of Quarks or leptons
Graviton/ Higg's Boson	weakest	Gravitational Force	(none)	1	(not known)	matter/ Higg's Field	the proximity of matter

Fermions

Fermions, since they conform to the Pauli Exclusion Principle, repulse each other and do not cluster; they can only be brought together by forces which override their inherent repulsion of each other. Fermions subdivide into two basic groups, termed **hadrons** (composed of **quarks**) and **leptons** (which are "pointal" and *not* further dividable.) Theoretically, at this point, neither quarks nor leptons would be capable of binding together into the units which we call atoms, due to their repulsion of each other. The Strong Force now enters the picture. Quarks *are* affected by the Strong Force (very strong, indeed, to overcome the Exclusion Principle), and they are bound together by this force into units called hadrons (e.g., protons and neutrons). Furthermore, the Strong Force not only binds quarks into hadrons, it also binds hadrons together into units which we have come to know as the nuclei of atoms. Leptons, inversely, *are not* affected by the Strong Force. They must, then, be completely true to the fermion portrait described by the Exclusion Principle and repulse each other fiercely.

This means that leptons, for example electrons, exhibit a powerful repulsion for their own kind. Thus negative electrons, while being tremendously attracted to the positive nucleus by the Electromagnetic Force, cannot overcome their repulsion for each other, and remain in an eternal "dance" around the nucleus.

Summary

The composite categorization of fermions and bosons set forth by Elementary Particle Physics provides more than a theoretical, more comprehensive system of classification with regard to matter and energy. It goes one step beyond the electrical and chemical reactions which the electronic structure of the atom explains, and brings us to the primal realm of explaining the very structure of the atom itself.

DR. LINDA BARTROM-OLSEN

CHAPTER 4

The Fermion Family

THE FERMION FAMILY (analogous to matter) family of particles are those subject to the Pauli Exclusion Principle. The two major sub-categories within this family are **leptons** and **hadrons**. Leptons are fermions which are "pointal" i.e. they cannot be further subdivided. Hadrons are fermions which are composed of still smaller particles called **quarks**.

Leptons

The lepton subdivision of the fermion family is characterized by three major properties:

1) Leptons are pointal,
2) Leptons are subject to the Pauli Exclusion Principle,
3) Leptons are not affected by the Strong Force, and therefore do not cluster.

The lepton family includes basically six particles and anti-particles. Lepton particles include the electron and e-neutrino, the muon and mu-neutrino, and the tau and tau-neutrino, shown in **Table 2**. The anti-leptons are the corresponding particles of identical mass but of opposite charges. All leptons and anti-leptons are considered to be elemental because (at the present time) it appears that they cannot be broken down into smaller entities, i.e. they give no hint of any internal structure.

General Lepton Classifications
—————— Table 2 ——————

MASSED LEPTONS	MASSLESS LEPTONS
electron	e-neutrino
positron ("anti-electron")	e-antineutrino
muon	μ- neutrino (mu-neutrino)
anti-muon	μ-antineutrino (mu-antineutrino)
tau	t-neutrino (tau-neutrino)
anti-tau	t-antineutrino (tau-antineutrino)

Since the leptons are generally found in motion (at incredibly high speeds), a further breakdown or classification of them is advanced dependent upon their spin (on their own individual axis) relative to the direction of their forward momentum. "Left-handed" is the term applied to all leptons which have an axial spin in the *opposite* direction of their momentum. Left-handed leptons include neutrinos, charged leptons (e.g., electrons), and positive muons. "Right-handed" leptons designate those leptons having spin in the same direction as their momentum and include anti-neutrinos, charged anti-leptons, and negative muons, **Table 3.**

DR. LINDA BARTROM-OLSEN

Selected Lepton Characteristics
——— Table 3 ———

Particle Name	Symbol	Rest Energy	Half-Life	Electrical Charge
e-neutrino	γ_e	0	Stable	0
e-antineutrino	$\bar{\gamma}_e$	0	Stable	0
μ-neutrino	γ_μ	0	Stable	0
μ-antineutrino	$\bar{\gamma}_\mu$	0	Stable	0
proton	e^+	0.51 MeV	Stable	+
electron	e^-	0.51 MeV	Stable	-
positive muon	μ^+	105.6 MeV	1.5×10^{-6} sec	+
negative muon	μ^-	105.6 MeV	1.5×10^{-6} sec	-

Hadrons

Hadrons, compared to leptons, are considerably more complex. There is evidence that they are not elemental but rather have some type of internal structure. More than 100 kinds of hadrons have been identified to this point! They include such particles as the proton, neutron, meson and hyperon, as well as their corresponding particles of anti-matter.

The great variety of hadrons, with their many differences, is now explained by a rather widely accepted theory of physics: that all hadrons are actually composed of a few simpler constituents called quarks. Six quarks have been identified, demonstrating the elegance of the parallel nature of matter-components. Although some properties of these quark units are lepton-like, there is an emphatic common denominator which fundamentally separates them from the leptons: their interactions are governed by the Strong Force, which does not affect leptons at all.

A theory such as this, which accounts for all the varieties of matter with just a few quarks (composing hadrons) and a few leptons, has an appealing economy. But although the Quark theory has gained widespread acceptance over the past decade, it must be acknowledged that there is no *direct* evidence that quarks themselves truly exist. That is, there is no direct evidence that they exist in isolation prior to hadrons being split by impact. Additionally, they are detected by devices specifically designed to note their presence. This fact in no way takes credence away from quark theory. But to neglect to point out these facts would be as great a logical error as it would be to dismiss the theory in the face of the quantity of information thus far accumulated. Quark *relations* between one another *are* experimentally observable, and so far they answer quantifiable questions better than any past premise has offered. They therefore provide a usable framework for our evolving portrait of what "matter" really is.

Quark Characteristics

The quark hypothesis was proposed independently in 1963 by Murray Gell-Mann and George Zweig, both of the California Institute of Technology. They named three types of quarks: the u-quark ("u" for up), the d-quark ("d" for down), and the s-quark ("s" for strange), as well as their anti-quark opposites: $\bar{u}$, $\bar{d}$ and $\bar{s}$. Two other physicists, Bjorken and Glashow of Harvard University named a fourth the "charmed quark" (designated as "c"), and the anti-matter particle, $\bar{c}$. In 1978, Leon Lederman contributed further through the discovery of a fifth quark which he designated as Upsilon. The Upsilon quark is more recently referred to as the bottom quark or b-quark. The sixth quark, the "top" quark, was discovered in 1995 and is the counterpart to the bottom quark. The u and d-quark are the smallest of the lot—an odd distinction since all quarks are nothing more than pointal in existence. While the charmed quark is comparatively massive, the Upsilon (or b) quark is the most massive of all. The s-quark uniquely displays the quality of "strangeness," a term applied to the degree of stability or resistance to decay, i.e. hadrons composed of s-quarks have very long lifetimes.

Quark Compositions

The entire hadron family is composed of these various quarks and their anti-matter counterparts in combinations which conform to three rules:

1. One possibility is for a quark and an anti-quark to bind together; the resulting particle is a member of the class of hadrons called **mesons** (e.g., a pi meson or pion is composed of a u-quark and a d-quark).
2. A second allowable combination consists of three quarks in a bound system. Hadrons formed in this way are called **baryons** (Greek for the "heavy one"). Examples include the proton (with the quark composition of uud) and the neutron (with the quark composition of udd).
3. The final (known) possible quark combination is the family of anti baryons, formed from three anti-quarks.

These are the only permissible ways, presently known, that combining quarks form hadrons. Obviously, other possible combinations do exist (such as particles of two quarks or of one quark and two anti-quarks) but such hadrons, at this time, are not known to exist.

The Quark Composition of Hadrons

Hadrons derive their properties **(see Table 4)** from the quarks which compose them, and a closer look at the possible properties which each quark may possess is now in order. Each quark, whether already known or yet awaiting discovery, transmits to the hadron it comprises three definitive characteristics: **baryon number, hypercharge**, and **electric charge.**

Baryon number involves the *number of quarks* composing the hadron. If the hadron is composed of two quarks, the hadron is a *meson* and has a baryon number of 0 (zero). If the hadron is made up of three quarks, the hadron is a *baryon* and has a baryon number of +1 or -1, depending on whether it is matter or anti-matter. Protons and neutrons, by way of example, are baryons. **Hypercharge** is a characteristic of hadrons involving the *cumulative spins of their composite quarks*. The spin of each quark is called its "aitch bar ($\hbar$) value." This characteristic designates, when all quarks composing a hadron are considered, the total orbital angular momentum of the hadron. The **electric charge** property of hadrons, assigned whole values, predetermines that the intrinsic electrical charge of an individual quark must be a fraction of the changes carried by all other known particles. Quarks, at this writing, are assigned electric charges of plus or minus 1/3 or multiples thereof; u-quarks are, for example, assigned an electric charge of +2/3, and d-quarks are assigned an electric charge of -1/3. Since protons are believed to be composed of the quark constituency uud, this results in a cumulative electric charge of (+2/3) + (+2/3) + (-1/3) = +1, which satisfies the present data on proton electric charge, shown in **Table 4.**

Selected Hadron Properties
————— Table 4 —————

	Hadron	Symbol	Rest Energy in MeV	Spin	Baryon Number
MESONS / 2q (even quark number)	neutral pion	π	135	0	0
	+ and - pions	π^+, π^-	140	0	0
	+ and - K mesons	K^+, K^-	494	0	0
	neutral K mesons	$\overset{\circ}{K}, \overline{K}^{\circ}$	498	0	0
Baryons / 3q (odd quark number)	proton & anti-proton	$p, \overline{p}$	938	$\frac{1}{2}\hbar$	+1, -1
	neutron & anti-neutron	$n, \overline{n}$	940	$\frac{1}{2}\hbar$	+1, -1
	lambda & anti-lambda	$\Lambda^{\circ}, \overline{\Lambda}^{\circ}$	1116	$\frac{1}{2}\hbar$	+1, -1
	sigma⁺ & anti-sigma⁺	$\Sigma^+, \overline{\Sigma}^-$	1189	$\frac{1}{2}\hbar$	+1, -1
	sigma° & anti-sigma°	$\Sigma^{\circ}, \overline{\Sigma}^{\circ}$	1192	$\frac{1}{2}\hbar$	+1, -1
	sigma⁻ & anti-sigma⁻	$\Sigma^-, \overline{\Sigma}^-$	1197	$\frac{1}{2}\hbar$	+1, -1
	xi° & anti-xi°	$\Xi^{\circ}, \overline{\Xi}^{\circ}$	1315	$\frac{1}{2}\hbar$	+1, -1
	xi⁻ & anti-xi⁻	$\Xi^-, \overline{\Xi}^-$	1321	$\frac{1}{2}\hbar$	+1, -1

The hadron subdivision of fermions is obviously more complicated than the subdivision of leptons since hadrons are composed of still smaller parts. Leptons, you will recall, are elemental. Again, hadron constituency is determined by the number and type of quarks which compose them. Therefore, the three basic characteristics of quarks (i.e. baryon number, hypercharge, and electric charge) predicates the major characteristic of hadrons, as well.

Baryon Number

Baryon Number is a property of hadrons dependent upon the number of quarks of which they are composed. There are two groups of hadrons designated according to their baryon number: **baryons** and **mesons**. The title of "baryon number" itself can be misleading; it was assigned because the *baryon subgroups* end up with a number (sometimes termed a conserved quantity) and *mesons* do *not*. **Baryons** are composed of three quarks and are therefore, *hadrons with a non-zero baryon number*. It has been proposed that baryons are comprised of two quarks (assigned +1 baryon number) and one anti-quark (assigned -1) For baryons this then yields a baryon number of +1+1-1= +1. Anti-baryons, on the other hand, are conjectured to be composed of two anti-quarks (assigned -1) and one quark (assigned +1), yielding a baryon number of -1-1+1= -1. With regard to this point, there appears, at this time, to be rather disparate conjecture at various locations of research-not an unfamiliar nor unexpected phenomenon at the beginning phase of any new branch of science.

Scientific findings often seem at odds with one another, and this certainly poses a problem relative to schools of thought conjecturing that quarks and anti-quarks do not combine. Nevertheless, it does serve as one major categorization of the hadron family. Protons and neutrons are examples of baryons. *Mesons*, it is conversely proposed, are composed of two quarks and are *hadrons with a baryon number of zero*. They have "rest energies" less than a proton and there is general agreement that they are composed of one quark and one anti-quark; this would yield a combination of +1-1=0, and indeed result in a baryon number of zero.

DR. LINDA BARTROM-OLSEN

Hypercharge

The second major classification of hadrons, according to the qualities of the quarks they contain, is **hypercharge**: a measure of orbital angular momentum or **spin**, sometimes called aitch bar value, and is represented by the symbol $\hbar$. Quarks are assigned spin values of plus or minus ½ depending on the direction of their spin. This characteristic, when referring to hadrons composed of three quarks, i.e. **baryons**, will always yield a *half integer spin*. Examples of combinations are:

$$+ \tfrac{1}{2} + \tfrac{1}{2} + \tfrac{1}{2} = 3/2 \ \hbar$$

Or

$$+ \tfrac{1}{2} - \tfrac{1}{2} + \tfrac{1}{2} = \tfrac{1}{2} \ \hbar$$

Mesons, on the other hand, composed of two quarks, can either have parallel spin:

$$+ \tfrac{1}{2} + \tfrac{1}{2} = 1 \hbar$$

Or opposite spin:

$$+ \tfrac{1}{2} - \tfrac{1}{2} = 0 \hbar$$

But in either case yield an *integer spin*.

Electric Charge

The third quality of hadrons is termed the **electric charge**. The electric charge is determined by the *type* of quarks of which a hadron is composed and is an *intrinsic quality of the quarks themselves*. This differentiates it from the baryon number, dependent upon the *number* of quarks present, or hypercharge which is dependent on their *orientation in space*. Since electric charges assigned to common hadrons are integers, it was decided that electric charges assigned to quarks must be fractions of charges carried by all other known particles. Four of the six conjectured quarks have been assigned electric values which conform to known baryon and meson electric charge values, **Table 5.** For protons and neutrons these electric charge values work out as follows:

Proton=

uud= +2/3e +2/3e -1/3e = +1e

Neutron=

udd= +2/3e -1/3e -1/3e = 0e

Summary

In summary, **fermions,** which comprise the portion of our universe known as **matter,** are subdivided into two categories: **leptons** and **hadrons. Leptons,** in accordance with the Pauli Exclusion Principle, do *not* cluster (i.e. they repulse each other). **Hadrons,** on the other hand, and the **quarks** which compose them, *will* cluster because the **Strong Force** overcomes their natural hadrons which pair or cluster in groups of *two* quarks are called **mesons.** And the hadrons clustering in groups of *three* quarks are called **baryons.**

The study of quark properties is sometimes called **quantum chromodynamics.** This terminology is derived from the expressions "flavor"—which designate whether a quark is up or down, strange or charmed, bottom or (tentatively named) top—and "color" which simply implies that there are further differences within each type of quark category. Each "flavor" of quark is thought to have three possible "colors" or further differences.

The main point remains, of course, that hadron classifications by quark quality (though beset with certain anomalies and conjectural divergences) is rapidly evolving into a structured, integrated scientific discipline. Ultimately, the discipline of Elementary Particle Physics is intended to simplify and classify the hundreds of sub-atomic particles into relatively few.

DR. LINDA BARTROM-OLSEN

Quark Classifications
Table 5

QUARK NAME	MATTER ABBREV.	ANTI MATTER ABBREV.	DISCOVERD BY	DATE	LOCATION	WEAK PEARING	ELECTRIC CHARGE (MATTER)
UP	u	$\overline{u}$	Gell - Mann and Zweig	1963	California Institute of Technology	u,d	$+2/3_e$
DOWN	d	$\overline{d}$	Gell - Mann and Zweig	1963	California Institute of Technology	d,u	$-1/3_e$
STRANGE	s	$\overline{s}$	Gell - Mann and Zweig	1963	California Institute of Technology	s,c	$-1/3_e$
CHARMED	c	$\overline{c}$	Bjorken and Glashow	1972	Harvard University	c,s	$+2/3_e$
UPSILON OR BOTTOM	b	$\overline{b}$	Lederman	1977	Fermilab	b,t	$-1/3_e$
"TOP"	t	$\overline{t}$	Abachi et al and Abe et al	1995	Fermilab and CERN	t,b	$+2/3_e$

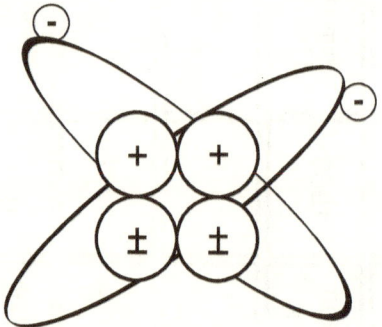

Pre Particle Physics
Leptons- $2e^-$
Hadrons- $2p^+$
          $2n^\pm$

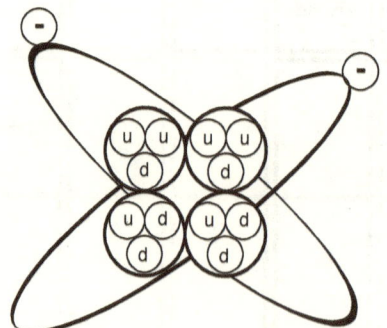

Post Particle Physics
Leptons- $2e^-$
Hadrons- $2p^+$ { Quarks; 2 up, 1 down
          $2n^\pm$ { Quarks;1 up, 2 down

Figure 7: A tale of Two Heliums

CHAPTER 5

The Boson Family: Forces and the Bosons which produce them

Forces

IN ADDITION TO the particle components of matter, *massless* particle components of energy are now recognized. These components of energy are called **bosons,** and bring about effects on, or changes in, matter, termed **forces.**

The forces, in order of the weakest through the strongest, are: the **Gravitational Force,** the **Weak Force,** the **Electromagnetic Force,** and the **Strong Force.** Generally speaking, the *greater* the mass of matter required to be present before the force is felt, the *weaker* the force is considered to be. Thus, the Gravitational Force, which acts primarily on grosser quantities of matter, is considered to be the weakest force; the Strong Force, on the other hand, which acts on quarks (the smallest component of matter), is considered to be the strongest force. The identifying characteristics of the Forces shown on **Table 6** are as follows:

The Gravitational Force

This force influences *all* particles in attracting them to each other, and its range is unlimited in a theoretical sense; its effect on sub-atomic particle, however, is negligible.

The Weak Force

The Weak Force is the second weakest force, but is still many orders of magnitude stronger than that of gravitation. It is still feeble enough to be observable only when the two strong forces are, for some reason, inhibited. Interactions where it is exhibited include: charged pion decay, strange hadrons, and among leptons, mesons, and baryons.

The Electromagnetic Force

The third force is the Electromagnetic Force. This force acts exclusively on particles that have an electric charge. Among these particles are two of the leptons (the electron and the muon), and all of the quarks. It is the electromagnetic force which binds atoms together, and is hence responsible for almost all of the gross properties of matter. This would include those designated as chemical, electrical and magnetic.

The Strong Force

The Strong Force underscores the difference between leptons and hadrons, or, according to the quark theory, between leptons and quarks. None of the leptons respond to the Strong Force. Only quarks (and hadrons which are composed of quarks) feel its influence. Quarks can interact with leptons through the weak and electromagnetic forces, but quarks interact *with each other* almost exclusively through the Strong Force. These interactions are what bind quarks together into protons and neutrons, while the electromagnetic Force binds the positive nucleus and negative electrons together into the structure which we call the atom. The Strong Force is more than one hundred times stronger than the Electromagnetic Force.

Characteristcs of Force
—— Table 6 ——

FORCE	RELATIVE STRENGTH	RANGE	PARTICLE(S) ACTED UPON	ACTIVATING BOSON
STRONG FORCE	1	Short Range	Quarks	Gluon
ELECTROMAGNETIC FORCE	10^{-2}	Long Range	Charged particles	Photon
WEAK FORCE	10^{-5}	Short Range	Quarks and Leptons	Intermediate vector boson
GRAVITATIONAL FORCE	10^{-39}	Long Range	All Particles	Graviton/ Higg's Boson

CHAPTER 6

Symmetry Theory of Forces

F ORCES EXPERIENCE A redefinition in Elementary
Particle Physics contrasted to their meaning in classical
Newtonian Physics, i.e. "a push or pull on an object." In
Elementary Particle Physics, forces tend to mean "an interaction
between." Exclusive of the Gravitational Force, which remains
something of an anomaly, forces rely for their definitive meaning on
what are called "Rules of Symmetry."

Overview

Other expressions for the four forces, which are useful in
studying Symmetry Theory, are:

1) The Electromagnetic Force is termed **QED** or
   Quantum Electro-Dynamics.
2) The Strong Force is termed **QCD** or
   Quantum Chromo-Dynamics.
3) The Weak Force is termed **QSD** or
   Quantum Spin-Dynamics.

The Terms QED, QCD, QSD in effect define the *type* of reaction
that the boson of each force is capable of bringing about on a
fermion. A more expansive way of looking at these three forces,
which at this writing have been defined with relative adequacy,
involves utilizing Symmetry Theory.

In Symmetry Theory, the boson involved in each type of force is examined in terms of a) *how many* particles it can affect at a time, and b) *what type of charge* a boson of that force is capable of rendering. The labeling which reflects the symmetry of a force involves a letter (or letters) followed by a number in parenthesis. The letter "S" stands for the sum of the weak charges of a doublet being zero; the letter "U" refers to the fact that what we are looking for is an overall unifying theory of forces; the number in parenthesis designates how many particles the boson in question may react with at any one time. Within the framework or terminology, the Electromagnetic Force of QED is considered to possess a U(1) symmetry; the Weak Force or QSD is portrayed as displaying an SU(2) symmetry; and the Strong Force or QCD is designated as having SU(3) symmetry.

Quantum Electro-Dynamics: U(1) Symmetry

The Electromagnetic Force is considered to possess a U(1) symmetry because the boson of QED, the photon, reacts with only one kind of particle at a time. It never transforms one kind of particle into another. The "U" designates that it is part of the unification theory, and there is no S because no particle is changed into another since a photon affects only one particle at a time.

Quantum Chromo-Dynamics: SU(3) Symmetry

The Strong Force or QCD is designated as having an SU(3) symmetry. Again, the "U" stands for the fact that we are looking for a unification of all forces: a fundamental way, that is, of relating one boson to another. The "3" stands for the fact that the Strong Force bosons, the gluons, are capable of transforming the three "colors" of quarks (an additional assigned quality of quarks) into one another. The "S" stands for the fact that the sum of the color charges (when three quarks are present) is zero; the quality of color charge, therefore, is generally undetectable in our universe as a whole, where matter (as we know it) is always composed of protons and neutrons having one of each color of quark present.

DR. LINDA BARTROM-OLSEN

Quantum Spin-Dynamics: SU(2) Symmetry

The Weak Force, or QSD, is portrayed as displaying an SU(2) symmetry. The "2" here stands for the fact that two members of what is termed a "doublet," can (through the Weak Force) be changed or "transformed" into each other. This is further explicated in the ensuing paragraphs illustrated in **Table 7**. The "U" reaffirms that the boson has been sufficiently described to be part of the unification theory. The "S" tells us that when the sum of the weak charges of a doublet are present, they effectively cancel each other out in terms of their detectability in gross amounts of matter.

Examples of Weak Doublets Include:
—————Table 7—————

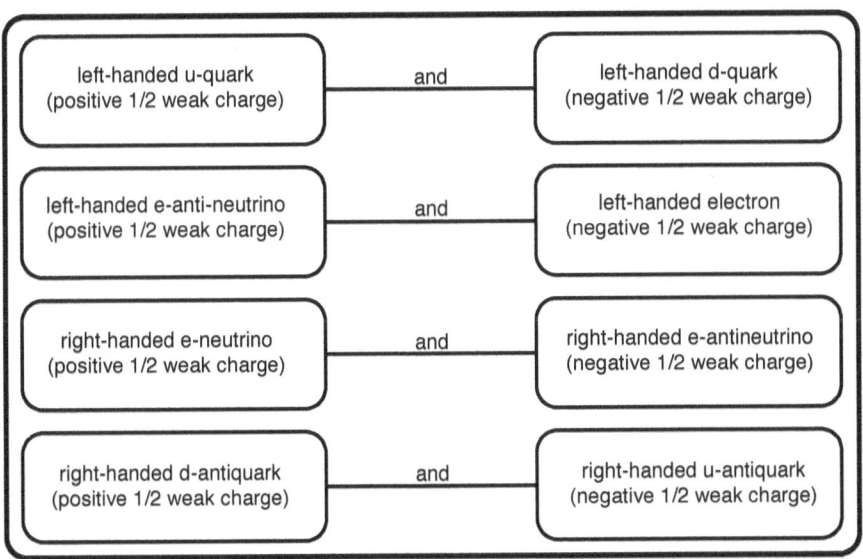

left-handed u-quark (positive 1/2 weak charge)	and	left-handed d-quark (negative 1/2 weak charge)
left-handed e-anti-neutrino (positive 1/2 weak charge)	and	left-handed electron (negative 1/2 weak charge)
right-handed e-neutrino (positive 1/2 weak charge)	and	right-handed e-antineutrino (negative 1/2 weak charge)
right-handed d-antiquark (positive 1/2 weak charge)	and	right-handed u-antiquark (negative 1/2 weak charge)

The Weak Force has remained something of an anomaly, while the Electromagnetic Force and the Strong Force have become more clearly defined over the last decade. Part of the reason for this may be that several of the reactions which the Weak Force explains have been traditionally incorporated under other headings of Physics such as *beta decay nuclear reaction,* **Table 8**. The Weak

Force portrait of transformation does not alter what is happening in these nuclear changes, but rather gives a more detailed explanation of each reaction, which appears to work better than theories which have existed in the past. One of the interesting attributes of the Weak Force is that only left-handed particles and right-handed anti-particles bear a weak charge.

Sample Weak Transformations
——————— Table 8 ———————

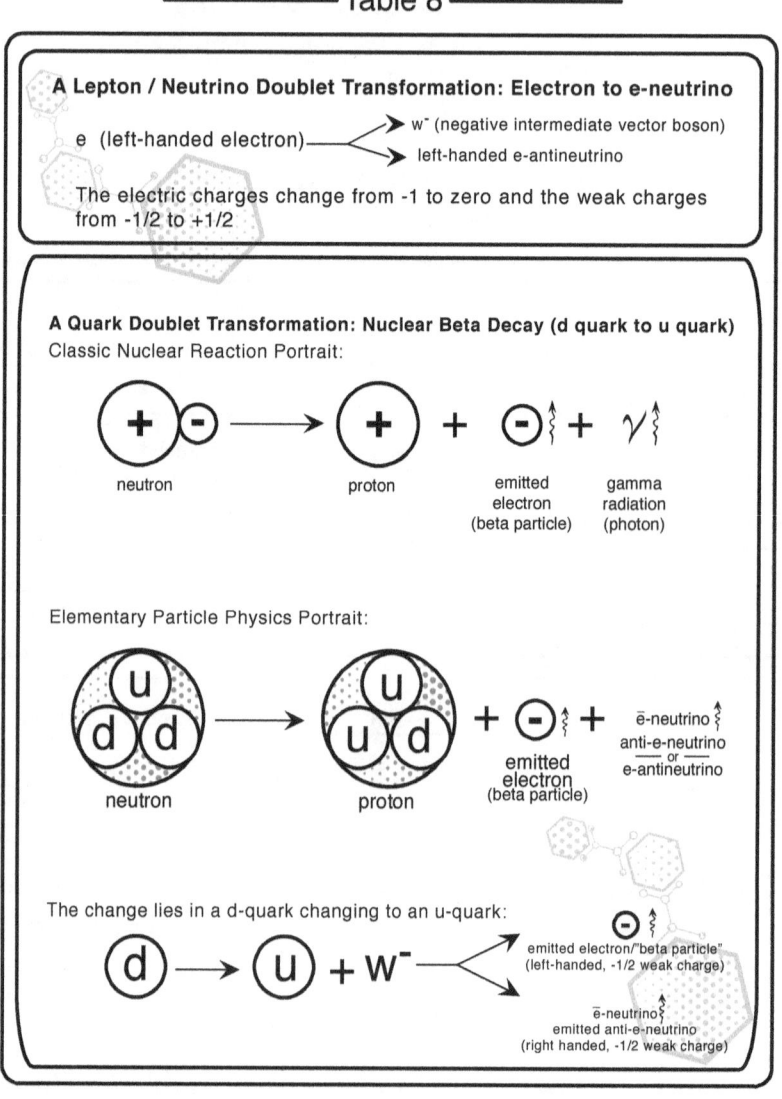

A Lepton / Neutrino Doublet Transformation: Electron to e-neutrino

e (left-handed electron) ——< w⁻ (negative intermediate vector boson)
left-handed e-antineutrino

The electric charges change from -1 to zero and the weak charges from -1/2 to +1/2

A Quark Doublet Transformation: Nuclear Beta Decay (d quark to u quark)
Classic Nuclear Reaction Portrait:

neutron → proton + emitted electron (beta particle) + gamma radiation (photon)

Elementary Particle Physics Portrait:

neutron → proton + emitted electron (beta particle) + ē-neutrino / anti-e-neutrino — or — e-antineutrino

The change lies in a d-quark changing to an u-quark:

(d) → (u) + w⁻ —< emitted electron/"beta particle" (left-handed, -1/2 weak charge)

ē-neutrino
emitted anti-e-neutrino
(right handed, -1/2 weak charge)

In summary, the Weak Force displays itself in terms of the intermediate vector bosons which are capable of transforming one member of a doublet into the other member.

Forces redefined in the context of Particle Physics

In studying the symmetry or likenesses that exist between forces, it might be wise to remind ourselves that in Elementary Particles Physics, "force" becomes simply a name for the ability to change the individual momenta, energy or identity of objects or particles. The units that are exchanged between particles of matter (fermions) are called bosons and are the units of force. One of the essential differences between forces is the *range* at which they have effect. For example, the Strong Force and the Weak Force are very different concerning the distances over which they affect particle. Reactions associated with the Strong Force have a 50% probability of occurring in matter that is 10^2 cm. (!!) of matter in thickness; the weak force, on the other hand, requires 10^{13} cm. (!!) of matter in thickness before there is a 50% probability of reaction. This is, in all likelihood, the reason that the terms "strong" and "weak" were applied to these reactions when they were named.

One of the basic scientific and philosophical functions of Elementary Particle Physics is, perhaps, to seek a unity between the types of forces and thereby secure a deeper understanding of the nature of our universe. One of the propositions that has been mad concerning such a unification was made in 1974 by Helen R. Quinn, who is now at the Stanford Linear Accelerator Center, by Steven Weinberg of Harvard University, and by Howard Georgi, also Professor of Physics at Harvard.

Unification extended: SU(5) Symmetry

What is proposed is a particle which can interchange between all three of the forces, QED, QSD and QCD, in order to provide an underlying unity among them. This particle is proposed to have a

SU(5) symmetry. The distance at which particles must be, in order to interchange and have this type of symmetry was determined to be 10^{-29} centimeters, and the particle was called an "X" particle. This distance is exceedingly small, but at this distance X particles can be exchanged, and is capable of transforming a quark into a lepton or a quark into an anti-quark. Theoretically, at this distance the Strong Force, the Weak Force, and the Electromagnetic Force become unified. Furthermore, there is a still smaller distance at which new phenomena are conjectured, incorporating Gravitational Force into the concept of unity. At about 10^{-33} centimeters, it has been suggested by some that gravitation may become as strong as the other three forces.

In order to push sub-subatomic particles to within 10^{-29} centimeters of each other requires strengths or energies far beyond the capabilities of man today. A distance of 10^{-29} centimeters in proximity corresponds to or requires an energy of about 10^{15} GeV (gigavolts), or roughly 10^{15} times the rest mass of a proton. If at this distance a quark can indeed be changed into a lepton, the energy required for this amount of compression of matter was probably present only at the beginning of the universe. This certainly holds no immediate danger to modern man. The fact remains, however, that the components of the atom as we know it *are* theoretically capable of breaking down. Our universe and ourselves could not then exist. The energies required to do so (or the amount of time involved) are, however, incredible. The present estimate is that the average life of a proton is about 10^{31} years. This is well beyond the expectations of life forms with which we are familiar on our planet!

This does, however, hold a clue as to the type of reactions that were going on in our universe when it was very, very new. In that intensely hot, incredibly compressed blue-white furnace of matter-energy, the cosmos as we know it today was born. In our ever-cooling universe, perhaps conjectures concerning our past can give rise to fascinating portraits of our future as well.

DR. LINDA BARTROM-OLSEN

CHAPTER 7

Symmetry Breaking

IN THE FIRST few seconds of the universe, there was matter-energy, with complete symmetry between the two. Within the first few microseconds of time, both the identity *between* and the symmetry *within* each category broke down as the material of the universe cooled. The general overview which follows is a simplistic introduction to what theories propose may have happened in the formation of matter-energy, i.e. fermions and bosons, as we know them today. It is a scenario of the beginning of time.

Fermions as a category first broke down or split into quarks and leptons. The major difference between them lay in the fact that leptons were unaffected by the Strong Force and *would not* cluster, whereas quarks *were* affected by the strong Force and *could* combine into units today called hadrons.

Next the quarks lost their similarity to each other and evolved into the separate entities being discovered today. The first generation included the up and down quarks, the second generation included the strange and charmed quarks, and the third generation included the upsilon and tau (or bottom and top) quarks.

The lepton category is also conjectured to have split into a family of different units in the first few seconds of existence. The first lepton generation is conjectured to have included the electrons (having mass) and the e-neutrinos (not having mass.) A second split in identity occurred with the second generation of leptons composed of the muon and the mu-neutrino. The third generation split produced the tau lepton and the tau-neutrino. Among the leptons, each generation included one massed and one massless particle.

Actually, the first generation of four fermions, the up and down quarks and the electron and e-neutrino leptons are all that is necessary to explain the composition of ordinary matter; the remaining second and third generation particles appear only in high energy experiments, even today.

At the beginning of time, all units of force (energy), the bosons, are conjectured to have been identical. At the tremendous energies present in the first few micro-seconds of time (set equal to a Planck mass which occurs at 10^{19} GeV), gravitational interactions of particles are thought to have had an important influence on their behavior. The largest particle accelerator built today has yet to attain 10^3 GeV. The conditions present at the beginning of our universe are, therefore, far from being capable of experimental reproduction.

The first split

The first split in the identity of bosons occurred as the universe cooled in the first seconds after the initial explosion; it was the division between the Gravitational Force and the after the initial explosion; it was the division between the Gravitational Force and the Hyper-Weak Force. (**See Table 9**) The intermediate vector baseball is the huge boson associated with Gravitation, and the smaller boson applies to the Hyper-Weak Force. It was still of high enough energy, however, to effectively change quarks to leptons or leptons to quarks. This is called the **Grand Unified Theory**, and occurs at energy levels of 10^{15} GeV. Although this is "low" enough in energy that Gravitation is a separate force, the other three forces (Strong, Electromagnetic, and Weak) are still identical, **Table 9**.

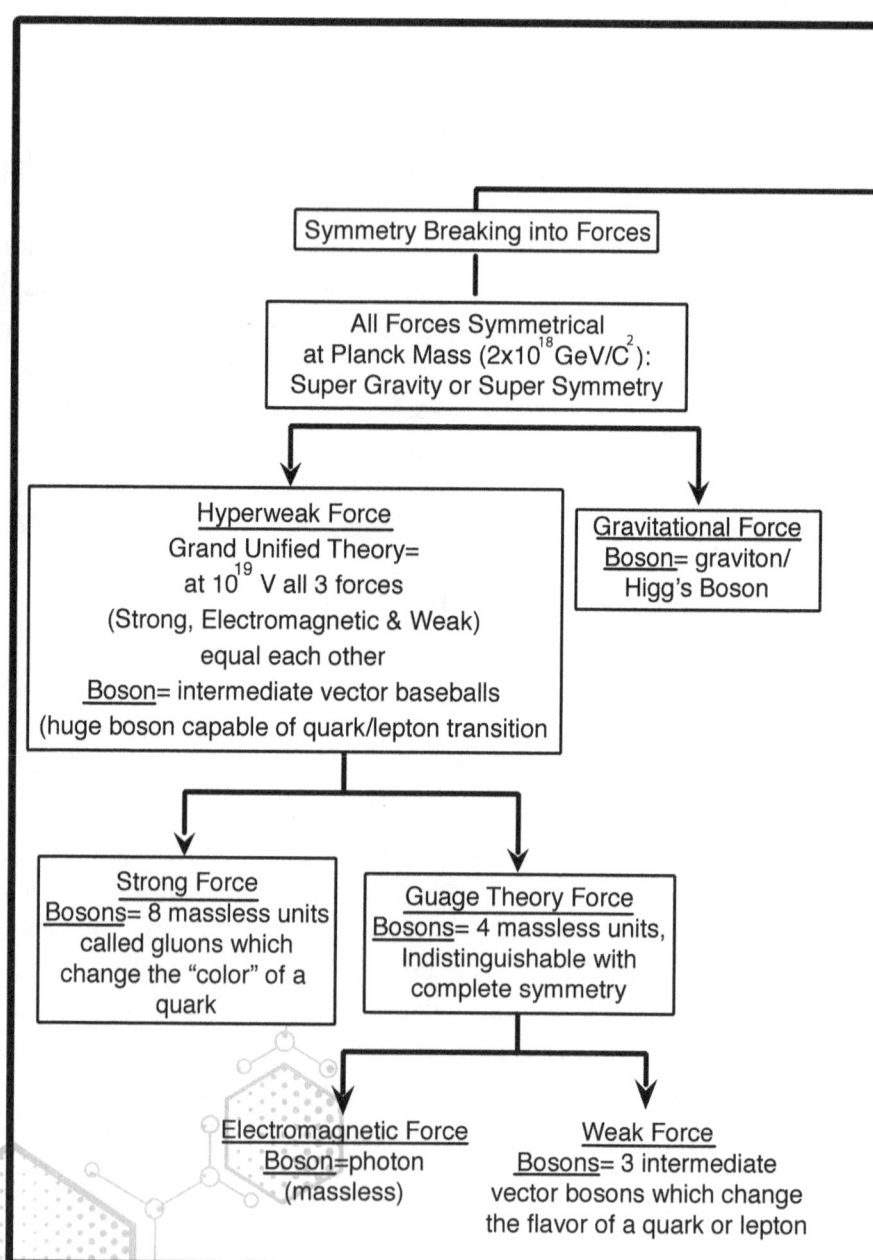

Symmetry Breaking into Forces

All Forces Symmetrical
at Planck Mass (2×10^{18} GeV/C^2):
Super Gravity or Super Symmetry

Hyperweak Force
Grand Unified Theory=
at 10^{19} V all 3 forces
(Strong, Electromagnetic & Weak)
equal each other
Boson= intermediate vector baseballs
(huge boson capable of quark/lepton transition

Gravitational Force
Boson= graviton/
Higg's Boson

Strong Force
Bosons= 8 massless units
called gluons which
change the "color" of a
quark

Guage Theory Force
Bosons= 4 massless units,
Indistinguishable with
complete symmetry

Electromagnetic Force
Boson=photon
(massless)

Weak Force
Bosons= 3 intermediate
vector bosons which change
the flavor of a quark or lepton

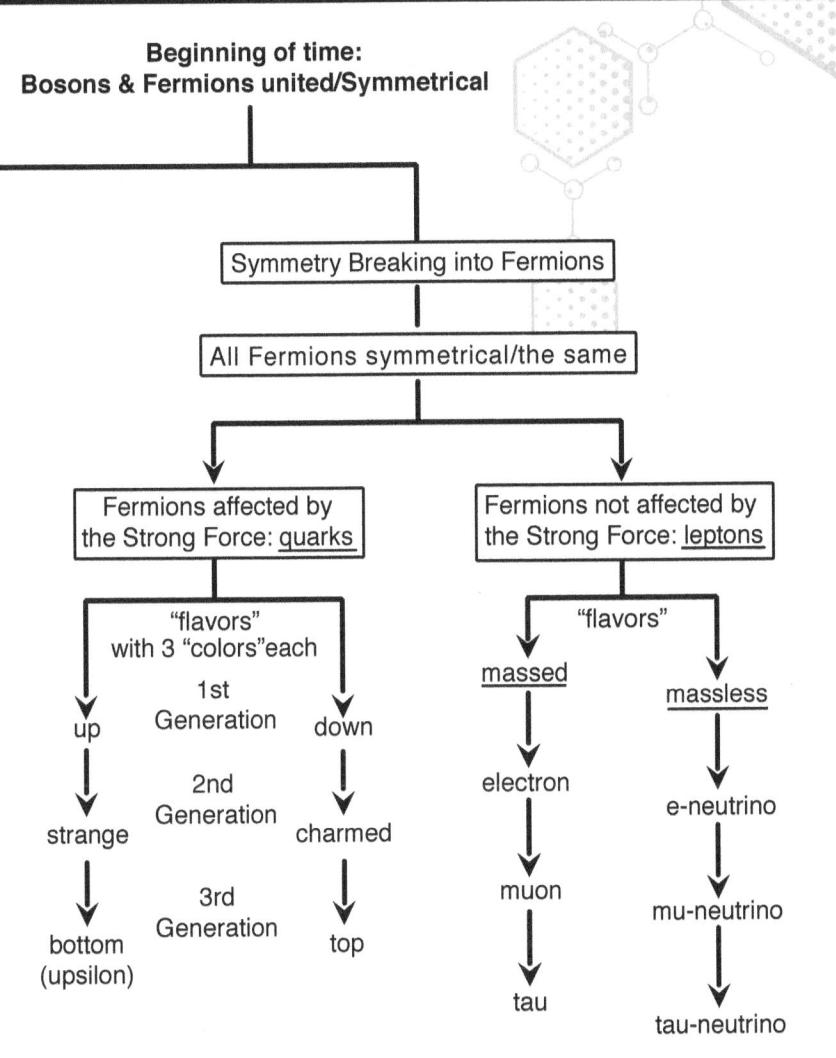

**Beginning of time:
Bosons & Fermions united/Symmetrical**

Symmetry Breaking into Fermions

All Fermions symmetrical/the same

Fermions affected by
the Strong Force: <u>quarks</u>

Fermions not affected by
the Strong Force: <u>leptons</u>

"flavors"
with 3 "colors"each

"flavors"

1st
Generation

up   down

<u>massed</u>

<u>massless</u>

2nd
Generation

electron

strange   charmed

e-neutrino

3rd
Generation

bottom   top
(upsilon)

muon

mu-neutrino

tau

tau-neutrino

The Second Split

The next break in symmetry between forces was when the Hyper-Weak Force subdivided into the Strong Force with eight mass- less bosons. At this point the quark/lepton subdivision becomes possible since a force now existed which affected *only* quarks. The Forces now were three, at only micro-seconds into existence of the universe: the Gravitational Force, the Strong Force, and the Gauge Theory Force.

The Final Split

The final break in symmetry or identity between the bosons of force occurred when the four identical, massless bosons of the Gauge Theory Force separated into the massless bosons of the Electromagnetic Force, the photon, and the massed bosons of the Weak Force, the intermediate vector bosons.

The laws of nature have, it appears, evolved from simpler state to the present complex one. The forces and the bosons which carry them out originally had a symmetry or similarity which was such that they could freely interchanged. It did not matter which force was applied, because the effect would be the same. Once symmetry was broken, this was no longer true. The differences among the forces which we observe today result from a process, just explained, called Spontaneous Symmetry Breaking. After symmetry was broken, the bosons each had different effects on fermions, and only one (the photon) remained massless.

Summary

In the beginning of time, the primitive forces themselves, at the massive energy level present at this time, had complete symmetry; leptons and quarks were interchangeable, symmetrical and massless. Even after the first split, the Strong, Weak, and Electromagnetic Forces were still unified. As the universal clock began, temperatures cooled, and energies plummeted into the comparatively "cool"

DR. LINDA BARTROM-OLSEN

universe of the present, "matter" and "energy" assumed their place in the cosmos.

Thus, in a breath of time, the material of the universe, "matter-energy/fermion-boson," lost mutual identity, broke symmetry, and subdivided into bosons and fermions and then further, into the more discrete units of which the galaxies and indeed our entire universe of today is composed.

CHAPTER 8

The Unification of Forces: An Historical Perspective

THE EFFORT BY physicists to develop new terminology and theory in the evolving picture of matter-energy is not an attempt to make the portrait of our universe more complex, but rather to make it simpler and clearer. When all units of both matter and energy are contained under an umbrella of a few quarks, leptons and bosons, the entire structure of our universe assumes a marvelous integrity. This integrity has a historical perspective while contributing to a scientific structure.

Kepler, Galileo and Newton

The first unification of any of the four forces occurred when the **Celestial Gravity Theory,** proposed by Kepler in the **Laws of Planetary Motion,** and **Terrestrial Gravity** put forth by Galileo, was unified under the **Gravitation Force** put forth by Isaac Newton in *Principia* published in 1687.

Maxwell

The separate phenomena of electricity, magnetism, and visible light were next perceived to be ramifications of the same force, and **were** unified under the title of Electromagnetism by Maxwell in his equations for the electromagnetic field during the 1800's.

Weinberg

Then, in 1968, Steven Weinberg of Harvard University proposed the Gauge Theory, which united the Electromagnetic Force with the Weak Force. This is considered by some to be particularly evident in the case of beta decay, in which a photon is evident as at least one type of boson involved (The boson of electromagnetism is the photon as well.)

Next there is a conjectured unification of Gauge Theory with the Strong Force, since the Intermediate Vector Bosons can influence both quarks and leptons in certain cases. Finally, the union of the Gauge and Strong Forces (termed the **Grand Unified Force**) could then unify with Newtonian gravity, and result in what is theoretically named **Supergravity, Table 10.**

Summary

These unifications, of which half provide the basis for modern physics while the other half provide, quite frankly, some rather esoteric conjecture, do provide undoubtedly one very valuable contribution to science. They move forward the ceaseless quest of the sciences to provide a final answer concerning what matter and energy (and the universe made of them) really are.

Celestial Gravity
(Kepler)

Terrestrial Gravity
(Galileo)

Electricity

Magnetism

Electromagnetic Force
Maxwell- equation for the
electromagnetic field-
19th Century

Light

Weak Force
(ex. Beta decay)

**Birth of
Elementary**
Particle Physics

Strong Force
(binds quarks and nuclei together)

of Forces

; of Physics)

e 10

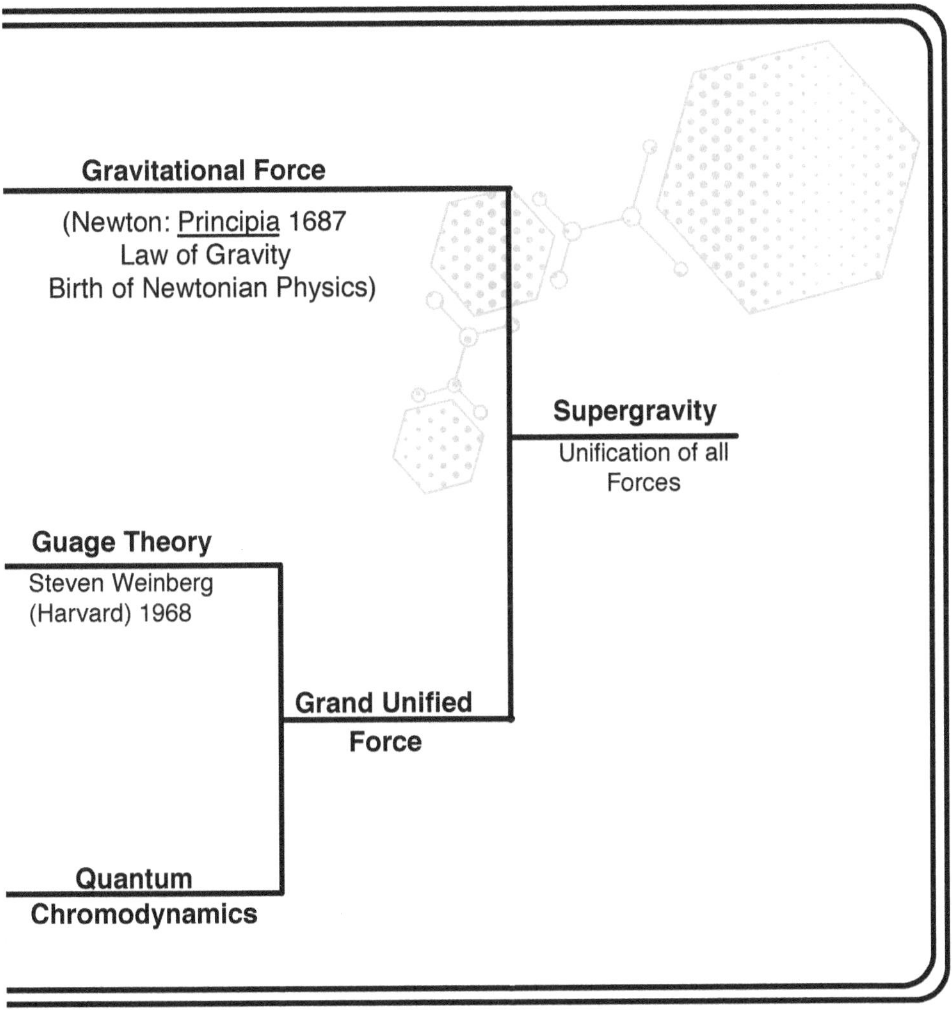

Gravitational Force
(Newton: Principia 1687
Law of Gravity
Birth of Newtonian Physics)
Supergravity
Unification of all
Forces
Guage Theory
Steven Weinberg
(Harvard) 1968
Grand Unified
Force
Quantum
Chromodynamics

CHAPTER 9

Summary

Bosons

THUS, IN THIS current portrait of the structure of matter-energy which comprises our universe, contemporary Elementary Particle Physics divides the components into Bosons and Fermions. Bosons, the quanta of energy (or forces), act upon Fermions which are the building blocks of matter.

Fermions

Fermions are categorized into leptons and hadrons which are in turn acted upon by four forces.

Leptons

The leptons are point-like in nature, and are acted upon by all of the forces except the strong force; examples of this family include the electron and the e-neutrino.

Hadrons

Hadrons, on the other hand, are composed of six still smaller units, called quarks. These quarks are bound together by the

Strong Force into members of the hadron family, two of which we commonly know as the proton and the neutron.

There are additional components of the quark theory, for example: the color and flavor of quarks, the almost certain possibility of the discovery of additional units, and the suggestion that the entire concept of "force" be replaced by the outlook of "interactions." This introduction is intended to provide a very basic literacy in Elementary Particle Physics, and through that a *foundation* for future exposure to this exciting, rapidly expanding field.

The implications of the lepton/hadron/forces prototype of sub-sub atomic particles are pervasive. Recent technological data provides both order and confirmable statistics concerning the relationship between the structure of matter and the forces that act upon it; the advent of the quark theory, in which most basic particles of our time-space continuum are viewed as having their prime difference in the *possession* of these forces, heralds the consequent restructuring of sub-atomic theory, unveiling worlds within worlds at the micro-cosmic level of our universe.

Echoes of the Past

PAVING THE WAY toward an understanding of the worlds within worlds which dwell in the land of particle physics, come forth the names which were the pioneers in the field, and with their names a clearer understanding of the origin of the particle names themselves. An echo of the past is embodied in their names; it bridges the particles of which we are made toward a future we cannot know, except fundamentally what we are made of. Physicists from Russia to California have given curious, magical and sometimes poetic names to the subatomic particles discovered over the last century or so, including their own names!

Beginning at the beginning with the symmetry splitting at the origin of our time continuum, comes the names of **boson** and **fermion**. **Bosons** (read particles of energy) are a class of particles often associated with forces (as the carriers of the force). They are named bosons because they obey Bose-Einstein statistics, named after the Indian physicist, **Satyendra Nath Bose** (1894-1974) who was a contemporary of Albert Einstein. The suffix "-on" is Greek, and became standard for newly discovered particles a century ago. **Fermions** (read particles of matter) are named for **Enrico Fermi**, Italian-born physicist (1901-1954), who built the first experimental nuclear reactor, the Atomic Pile of the University of Chicago; also named for Fermi is *Fermilab*, 35 mi west of Chicago. The Fermi National Accelerator Laboratory is America's particle physics laboratory.

Fermions are subdivided into **Hadrons** and **Leptons**. These names were proposed by the Russian theoretical physicist Lev Okun in 1962 when he wrote: "In this report I shall call strongly interacting particles 'hadrons' . . . [since] the Greek **hadros** signifies "large", [or] "massive", in contrast to **leptos** which means "small", [or] "light" [or thin]. I hope that this terminology will prove to be convenient."

Hadrons are composed of still smaller particles called **quarks**, named by American physicist Murray Gell-Mann in 1962. He had already come up with the sound, and was thinking of spelling it "kwork". He says in his book, the *Quark and the Jaguar:* "Then, in one of my occasional perusals of *Finnegans Wake*, by James Joyce, I came across the word 'quark' in the phrase 'Three quarks for Muster Mark'," and the name was born! Quarks compose protons and neutrons in the atomic nucleus, as well as many other fermions. Ernest Rutherford in 1920 named the **Proton**, the hydrogen nucleus, from the Greek "protos" meaning "first" since Hydrogen is the first element. Both protons and neutrons are made of three quarks. **Mesons** are particles made of two quarks: a quark and an anti-quark. That name comes from the Greek "meso" meaning "mid", because mesons, when first observed, appeared to have a mass somewhere between that of an electron, and nucleons (protons and neutrons).

Leptons in contrast are not composed of smaller parts, they are "pointal", and include **electrons** and **neutrinos**. An **electron,** part of a normal atom which orbits the nucleus, is an indivisible quantity of electric charge, proposed in 1894 by the Irish physicist, **George Johnston Stoney** (1826-1911), derived from the word "electric" (or the Latin "electrum") plus the Greek suffix "-on". The **neutrino** was named by **Enrico Fermi** who gave the neutrino an Italian name which means "small neutral one" because it has such a tiny mass even by the standards of subatomic particles. It was actually first named "neutron" by Wolfgang Pauli (1900-1958) in 1930, but Fermi renamed it three years later, because "neutron" (from the Latin for "neutral") had by then begun to be used to refer to the uncharged particle in an atom's nucleus.

Bosons, particles of energy, include the **Gluon**, a type of boson responsible for the strong force between quarks. The term derives from the English word "glue" and was first proposed in 1962 by **Gell-Mann**, who also suggested the existence of particles actually composed of a number of gluons, which he called "glueballs". Another boson, the **Photon**, is a name derived from the Greek "phos" meaning "light." The **Higgs boson** is named for British physicist Peter Higgs, one of the first to propose its existence in 1964. It has also been named "The God particle" by American physicist **Leon Lederman**. "He wanted to refer to it as that 'goddamn particle' and his editor wouldn't let him," Higgs told the *Guardian*. So "God particle" it was; much magical and profound conjecture has surrounded this boson by the media due to the very power of its name, when in truth it was a penstroke of editing that named it!

So these are a few of the stories which surround the names of elementary particles and through which they are called up and carried into the future. Such is the power of the past, of dreams long ago, of physicists long passed, whose spirits live on through the dedication of the men and women of science: those who use their names, their knowledge, their spirit-power, to arch us into our future understanding of the fundamental stuff of our time-continuum and the elemental wisps of particles of which we are made.

REVIEW QUESTIONS

1. What are the parts of an atom? How are they different?
2. What keeps electrons in orbit around the nucleus?
3. What did Newlands notice that led him to his Table of Elements?
4. Mendeleev's Periodic System left spaces empty. Why?
5. About how many elements do we know of now?
6. Are most elements Metals, Non-Metals, or Inert Gases?
7. What is the difference between a Boson and a Fermion?
8. Name the four major Forces recognized in Elementary Particle Physics.
9. How does the Strong Force effectively subdivide Fermions into two subgroups, leptons and hadrons?
10. Explain the difference between the two hadron families, baryons and mesons, in terms of the number of quarks each contains.
11. Discuss the Heisenberg Uncertainty Principle.
12. Name and briefly discuss the three main properties which a quark may possess.
13. Is the concept of unifying forces a new idea? Give a specific example to support your answer.
14. How does the Pauli Exclusion Principle combine with the Strong Force Theory to differentiate leptons from hadrons?
15. Scientists:

   - Who proposed that every eight elements act alike?
   - Who is credited with inventing the Periodic Table?
   - Who proposed the quark theory in 1963?
   - Who named the charmed quark?

- At what university did the scientist work who named the c quark?

16. What two forces does the Guage Theory unite, who is its proponent, and from which university does his work originate?
17. How did the name "quark" originate?

GLOSSARY OF TERMS

1. **Atom**—basic unit of matter, over 100 discovered or made, composed of a nucleus of protons and neutrons and orbited by electrons.
2. **Baryon**—A hadron composed of three quarks.
3. **Boson**—Unit analogous to energy, which is interchanged among electrically charged fermions. Does not conform to the Pauli Exclusion Principle.
4. **Electromagnetic Force**—Causes interactions among electrically charged fermions.
5. **Electron**—a lepton; negative sub-atomic particle which orbits the nucleus of an atom
6. **Expanded Periodic Table**—fits all elements onto one table
7. **Fermion**—A unit analogous to matter, which interacts with other fermions by exchanging bosons. Conforms to the Pauli Exclusion Principle.
8. **Force**—Observed phenomenon of objects changing momentum and energy when near each other.
9. **Gluon**—Boson of the Strong Force.
10. **Gravitational Force**—Attracts gross amounts of fermions to other fermions.
11. **Graviton**—Boson of the Gravitational Force.
12. **Hadron**—A fermion which is made of smaller particles called quarks.
13. **Intermediate Vector Boson**—Boson of the Weak Force.
14. **Law of Octaves**—states that every seven elements repeat characteristics
15. **Lepton**—Fermion which is pointal, not composed of smaller particles.
16. **Meson**—A hadron composed of two quarks.
17. **Nucleus**—Center of an atom made of protons and neutrons

18. **Neutron**—neutral particle located in the nucleus of an atom
19. **Periodic Table of the Elements**—graphic which organizes all known elements to show their sizes and their characteristics
20. **Photon**—Boson of the Electromagnetic Force.
21. **Proton**—positive sub-atomic particle located in the nucleus of an atom
22. **Quark**—Unit composing fermions.
23. **Strong Force**—Binds quarks together into hadrons.
24. **Weak Force**—Changes quark or lepton paired units into each other.

REFERENCES

Abbott, Larry. "The Mystery of the Cosmological Constant," *Scientific American*, Vol. 258, #5, pp. 106-113

Adair, Robert. *The Great Design: Particles, Fields, and Creation.* Oxford University Press, 1987.

Adair, Robert. "A Flaw in the Universal Mirror," *Scientific American*, Vol. 258, #2, pp. 50-56.

Bartrom, L. et al. (1983) *School Science and Mathematics.*

Clark, R.C. & Lyons, C. (2011) *Graphics for Learning*—2nd Edition.

Feinberg, Gerald. *What is the World Made Of?* Anchor Press, New York, 1978

Feynman, Richard P. *Elementary Particles and the Laws of Physics: The 1986 Dirac Memorial Lectures.* Cambridge University Press, 1987.

Freedman, Daniel Z. and Peter van Nieuwenhuizen. "Supergravity and the Unification of the Law of Physics," *Scientific American*, Vol. 238, #2, pp. 126-143.

Frey, Raymond. *Elementary Particles*, lecture series. University of Oregon, 2013.

Gaillard, Marg K., Benjamin W. Lee and Jonathan L. Rosner. "Search for Charm," *Review of Modern Physics.* Vol. 47, #2, pp. 277-298

Georgi, Howard. "A Unified Theory of Elementary Particles and Forces," *Scientific American.* Vol. 258, #3, pp. 48-56.

Higgs Boson (2013) *http://home.web.cern.ch/about/physics/search-higgs-boson*

Hirsch, M., Päs, H., and Porad, Werner. "Ghostly Beacons of the New Physics", *Scientific American*, April 2013. pp 40-47.

Hoofti, Gerard. "Guage Theories of the Forces Between Elementary Particles," *Scientific American*, Vol. 242, #6, pp. 104-141

Jaffe, R. L. "Quark Confinement," *Nature,* Vol. 268 pp. 201-208

Kamenluchi, S., H. Ezawa, Y. Murayama, M. Namiki, S. Nomura, Y. Ohnuki And T. Yajima. *Foundations of Quantum Mechanics in the Light of New Technology.* Physical Society of Japan, 1984.

Krauss, Frank. *Introduction to Particle Physics,* lecture series. University of Durham, Stockton, England, Epiphany Term, 2010.

Lederman, Leon M. "The Upsilon Particle," *Scientific American,* Vol. 239, #4, pp. 72-103.

Mendeleev, Dmitri. (1869) *Zeitschrift für Chemie.*

Newlands, John A. R. (1865) *Chemical News.*

Ohanian, Hans C. *Gravitation and Spacetime,* W. W. Norton and Co., 1976.

Polkinghome, J. C. *The Quantum World.* Princeton University Press, 1985.

Rae, Alastair I. M. *Quantum Physics: Illusion or Reality?* Cambridge University Press, 1986

Schwitters, Roy. "Fundamental Particles with Charm," *Scientific American,* Vol. 237, #4, pp. 56-83

Shimony, Abner. "The Reality of the Quantum World," *Scientific American,* Vol. 258, #1, pp. 46-53

Standard Model of Particle Physics (2012) *http://physics.info/standard/*

Vee Mapping with Junior High School Science Students. Science Education (p. 625).

Wilczek, Frank. *The Physics of Nothing. http://www.pbs.org/wgbh/nova/physics/* blog/author/fwilczek/

Zukav, Gary. *The Dancing Wu Li Masters: An Overview of the New Physics.* Wm. Morrow and Co., New York, 1979.

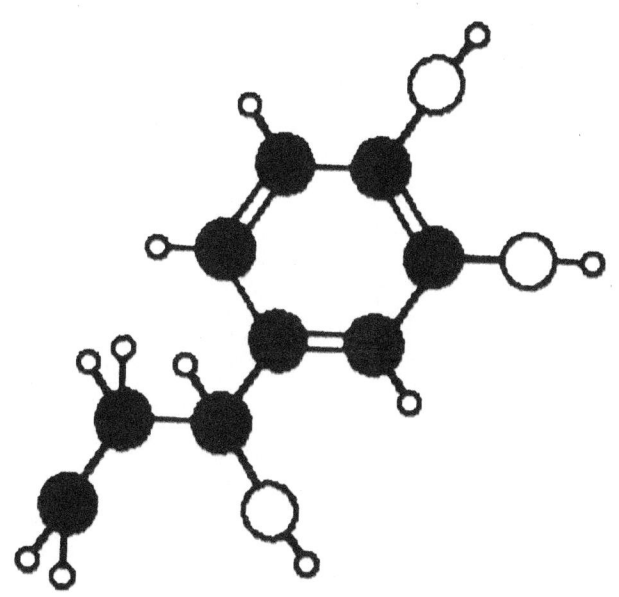

DR. LINDA BARTROM-OLSEN

DR. LINDA BARTROM-OLSEN

www.ingramcontent.com/pod-product-compliance
Lightning Source LLC
Chambersburg PA
CBHW021016180526
45163CB00005B/1975